THIS BOOK BELONGS TO

EMAIL:

ADDRESS:

CONTACT:

PHONE:

START DATE	END DATE

MO TU WE TH FR SA SU
☐   ☐   ☐   ☐   ☐   ☐   ☐

DATE: ___ / ___ / ___

PROJECT: _____

FOREMAN: _____

WEATHER   F°____ C°____   ____AM ____PM

HOURS DUE TO BAD WEATHER	ISSUED AND DELAYS

NOTE: _____

COMPLETION DATE	DAYS AHEAD OF SCHEDULE	DAYS BEHIND SCHEDULE

SAFETY AND INCIDENTS

SAFETY ISSUES THAT NEED TO BE ADDRESSED	ACCIDENTS / INCIDENTS / STEPS NEEDED TO RESOLVE

SUMMARY OF THE WORK DONE TODAY

IMPORTANT NOTES

NAME	SIGNATURE

TODAY LABOR

INITIALS	TRADE	START	FINISH	PAID HOURS	OVERTIME	COMPANY
☐ EMPLOYEE ☐ CONTRUCTOR		AM	PM			
☐ EMPLOYEE ☐ CONTRUCTOR		AM	PM			
☐ EMPLOYEE ☐ CONTRUCTOR		AM	PM			
☐ EMPLOYEE ☐ CONTRUCTOR		AM	PM			
☐ EMPLOYEE ☐ CONTRUCTOR		AM	PM			
☐ EMPLOYEE ☐ CONTRUCTOR		AM	PM			
☐ EMPLOYEE ☐ CONTRUCTOR		AM	PM			
☐ EMPLOYEE ☐ CONTRUCTOR		AM	PM			

EQUIPMENT ON SITE	NO. OF UNITE	WORKING YES / NO

HIRED EQUIPMENT	NO. OF UNITE	EQUIPMENT RENTED	FROM	RATE

NAME: _____   SIGNATURE: _____

MO TU WE TH FR SA SU
☐   ☐   ☐   ☐   ☐   ☐   ☐

DATE: /   /

PROJECT:

FOREMAN:

WEATHER

F°   C°   AM   PM

| HOURS DUE TO BAD WEATHER | ISSUED AND DELAYS |

NOTE: _____

COMPLETION DATE	DAYS AHEAD OF SCHEDULE	DAYS BEHIND SCHEDULE

SAFETY AND INCIDENTS

SAFETY ISSUES THAT NEED TO BE ADDRESSED	ACCIDENTS / INCIDENTS / STEPS NEEDED TO RESOLVE

SUMMARY OF THE WORK DONE TODAY

IMPORTANT NOTES

NAME	SIGNATURE

TODAY LABOR

INITIALS	TRADE	START	FINISH	PAID HOURS	OVERTIME	COMPANY
☐ EMPLOYEE ☐ CONTRUCTOR		AM	PM			
☐ EMPLOYEE ☐ CONTRUCTOR		AM	PM			
☐ EMPLOYEE ☐ CONTRUCTOR		AM	PM			
☐ EMPLOYEE ☐ CONTRUCTOR		AM	PM			
☐ EMPLOYEE ☐ CONTRUCTOR		AM	PM			
☐ EMPLOYEE ☐ CONTRUCTOR		AM	PM			
☐ EMPLOYEE ☐ CONTRUCTOR		AM	PM			
☐ EMPLOYEE ☐ CONTRUCTOR		AM	PM			

EQUIPMENT ON SITE	NO. OF UNITE	WORKING YES / NO

HIRED EQUIPMENT	NO. OF UNITE	EQUIPMENT RENTED	FROM	RATE

NAME: _____   SIGNATURE: _____

MO	TU	WE	TH	FR	SA	SU		DATE:	/	/
:-:	:-:	:-:	:-:	:-:	:-:	:-:				
☐	☐	☐	☐	☐	☐	☐				

PROJECT: **FOREMAN:**

WEATHER

F°_____ C°_____ _____ AM _____ PM

HOURS DUE TO BAD WEATHER	ISSUED AND DELAYS

NOTE: _____

COMPLETION DATE	DAYS AHEAD OF SCHEDULE	DAYS BEHIND SCHEDULE

SAFETY AND INCIDENTS

SAFETY ISSUES THAT NEED TO BE ADDRESSED	ACCIDENTS / INCIDENTS / STEPS NEEDED TO RESOLVE

SUMMARY OF THE WORK DONE TODAY

IMPORTANT NOTES

NAME	SIGNATURE

TODAY LABOR

INITIALS	TRADE	START	FINISH	PAID HOURS	OVERTIME	COMPANY
☐ EMPLOYEE ☐ CONTRUCTOR		AM	PM			
☐ EMPLOYEE ☐ CONTRUCTOR		AM	PM			
☐ EMPLOYEE ☐ CONTRUCTOR		AM	PM			
☐ EMPLOYEE ☐ CONTRUCTOR		AM	PM			
☐ EMPLOYEE ☐ CONTRUCTOR		AM	PM			
☐ EMPLOYEE ☐ CONTRUCTOR		AM	PM			
☐ EMPLOYEE ☐ CONTRUCTOR		AM	PM			
☐ EMPLOYEE ☐ CONTRUCTOR		AM	PM			

EQUIPMENT ON SITE	NO. OF UNITE	WORKING YES / NO

HIRED EQUIPMENT	NO. OF UNITE	EQUIPMENT RENTED	FROM	RATE

NAME: _____   SIGNATURE: _____

MO   TU   WE   TH   FR   SA   SU
☐   ☐   ☐   ☐   ☐   ☐   ☐

DATE: ___ / ___ / ___

PROJECT: _____

FOREMAN: _____

WEATHER ☁️ ⛅ ☁️ 🌨️ ☀️ 🌧️ ⛈️

F°____ C°____ _____ AM _____ PM

HOURS DUE TO BAD WEATHER	ISSUED AND DELAYS

NOTE: _____

COMPLETION DATE	DAYS AHEAD OF SCHEDULE	DAYS BEHIND SCHEDULE

SAFETY AND INCIDENTS

SAFETY ISSUES THAT NEED TO BE ADDRESSED	ACCIDENTS / INCIDENTS / STEPS NEEDED TO RESOLVE

SUMMARY OF THE WORK DONE TODAY

_____
_____
_____
_____
_____
_____

IMPORTANT NOTES

_____
_____
_____
_____
_____
_____

NAME	SIGNATURE

TODAY LABOR

INITIALS	TRADE	START	FINISH	PAID HOURS	OVERTIME	COMPANY
☐ EMPLOYEE ☐ CONTRUCTOR		AM	PM			
☐ EMPLOYEE ☐ CONTRUCTOR		AM	PM			
☐ EMPLOYEE ☐ CONTRUCTOR		AM	PM			
☐ EMPLOYEE ☐ CONTRUCTOR		AM	PM			
☐ EMPLOYEE ☐ CONTRUCTOR		AM	PM			
☐ EMPLOYEE ☐ CONTRUCTOR		AM	PM			
☐ EMPLOYEE ☐ CONTRUCTOR		AM	PM			
☐ EMPLOYEE ☐ CONTRUCTOR		AM	PM			

EQUIPMENT ON SITE	NO. OF UNITE	WORKING YES / NO

HIRED EQUIPMENT	NO. OF UNITE	EQUIPMENT RENTED	FROM	RATE

NAME: _____   SIGNATURE: _____

MO TU WE TH FR SA SU
☐   ☐   ☐   ☐   ☐   ☐   ☐

DATE: /   /

PROJECT:

FOREMAN:

WEATHER

F°___ C°___   ___ AM ___ PM

HOURS DUE TO BAD WEATHER	ISSUED AND DELAYS

NOTE: _____

COMPLETION DATE	DAYS AHEAD OF SCHEDULE	DAYS BEHIND SCHEDULE

SAFETY AND INCIDENTS

SAFETY ISSUES THAT NEED TO BE ADDRESSED	ACCIDENTS / INCIDENTS / STEPS NEEDED TO RESOLVE

SUMMARY OF THE WORK DONE TODAY

IMPORTANT NOTES

NAME	SIGNATURE

TODAY LABOR

INITIALS	TRADE	START	FINISH	PAID HOURS	OVERTIME	COMPANY
☐ EMPLOYEE ☐ CONTRUCTOR		AM	PM			
☐ EMPLOYEE ☐ CONTRUCTOR		AM	PM			
☐ EMPLOYEE ☐ CONTRUCTOR		AM	PM			
☐ EMPLOYEE ☐ CONTRUCTOR		AM	PM			
☐ EMPLOYEE ☐ CONTRUCTOR		AM	PM			
☐ EMPLOYEE ☐ CONTRUCTOR		AM	PM			
☐ EMPLOYEE ☐ CONTRUCTOR		AM	PM			
☐ EMPLOYEE ☐ CONTRUCTOR		AM	PM			

EQUIPMENT ON SITE	NO. OF UNITE	WORKING YES / NO

HIRED EQUIPMENT	NO. OF UNITE	EQUIPMENT RENTED	FROM	RATE

NAME: _____   SIGNATURE: _____

MO TU WE TH FR SA SU
☐   ☐   ☐   ☐   ☐   ☐   ☐

DATE: /   /

PROJECT:

FOREMAN:

WEATHER

F°____   C°____   ____AM   ____PM

| HOURS DUE TO BAD WEATHER | ISSUED AND DELAYS |

NOTE: _____

COMPLETION DATE	DAYS AHEAD OF SCHEDULE	DAYS BEHIND SCHEDULE

SAFETY AND INCIDENTS

SAFETY ISSUES THAT NEED TO BE ADDRESSED	ACCIDENTS / INCIDENTS / STEPS NEEDED TO RESOLVE

SUMMARY OF THE WORK DONE TODAY

IMPORTANT NOTES

NAME	SIGNATURE

TODAY LABOR

INITIALS	TRADE	START	FINISH	PAID HOURS	OVERTIME	COMPANY
☐ EMPLOYEE ☐ CONTRUCTOR		AM	PM			
☐ EMPLOYEE ☐ CONTRUCTOR		AM	PM			
☐ EMPLOYEE ☐ CONTRUCTOR		AM	PM			
☐ EMPLOYEE ☐ CONTRUCTOR		AM	PM			
☐ EMPLOYEE ☐ CONTRUCTOR		AM	PM			
☐ EMPLOYEE ☐ CONTRUCTOR		AM	PM			
☐ EMPLOYEE ☐ CONTRUCTOR		AM	PM			
☐ EMPLOYEE ☐ CONTRUCTOR		AM	PM			

EQUIPMENT ON SITE	NO. OF UNITE	WORKING YES / NO

HIRED EQUIPMENT	NO. OF UNITE	EQUIPMENT RENTED	FROM	RATE

NAME: _____   SIGNATURE: _____

MO TU WE TH FR SA SU
☐   ☐   ☐   ☐   ☐   ☐   ☐

DATE: ___ / ___ / ___

PROJECT: _____

FOREMAN: _____

WEATHER   ☁ ⛅ ☁ 🌨 ☀ 🌧 ⛈

F° ____ C° ____ ____ AM ____ PM

HOURS DUE TO BAD WEATHER	ISSUED AND DELAYS

NOTE: _____

COMPLETION DATE	DAYS AHEAD OF SCHEDULE	DAYS BEHIND SCHEDULE

SAFETY AND INCIDENTS

SAFETY ISSUES THAT NEED TO BE ADDRESSED	ACCIDENTS / INCIDENTS / STEPS NEEDED TO RESOLVE

SUMMARY OF THE WORK DONE TODAY

IMPORTANT NOTES

NAME	SIGNATURE

TODAY LABOR

INITIALS	TRADE	START	FINISH	PAID HOURS	OVERTIME	COMPANY
☐ EMPLOYEE ☐ CONTRUCTOR		AM	PM			
☐ EMPLOYEE ☐ CONTRUCTOR		AM	PM			
☐ EMPLOYEE ☐ CONTRUCTOR		AM	PM			
☐ EMPLOYEE ☐ CONTRUCTOR		AM	PM			
☐ EMPLOYEE ☐ CONTRUCTOR		AM	PM			
☐ EMPLOYEE ☐ CONTRUCTOR		AM	PM			
☐ EMPLOYEE ☐ CONTRUCTOR		AM	PM			
☐ EMPLOYEE ☐ CONTRUCTOR		AM	PM			

EQUIPMENT ON SITE	NO. OF UNITE	WORKING YES / NO

HIRED EQUIPMENT	NO. OF UNITE	EQUIPMENT RENTED	FROM	RATE

NAME: _____   SIGNATURE: _____

MO TU WE TH FR SA SU
☐   ☐   ☐   ☐   ☐   ☐   ☐

DATE: /   /

PROJECT:

FOREMAN:

WEATHER

F°   C°   ___AM   ___PM

HOURS DUE TO BAD WEATHER	ISSUED AND DELAYS

NOTE: _____

COMPLETION DATE	DAYS AHEAD OF SCHEDULE	DAYS BEHIND SCHEDULE

SAFETY AND INCIDENTS

SAFETY ISSUES THAT NEED TO BE ADDRESSED	ACCIDENTS / INCIDENTS / STEPS NEEDED TO RESOLVE

SUMMARY OF THE WORK DONE TODAY

IMPORTANT NOTES

NAME	SIGNATURE

TODAY LABOR

INITIALS	TRADE	START	FINISH	PAID HOURS	OVERTIME	COMPANY
☐ EMPLOYEE ☐ CONTRUCTOR		AM	PM			
☐ EMPLOYEE ☐ CONTRUCTOR		AM	PM			
☐ EMPLOYEE ☐ CONTRUCTOR		AM	PM			
☐ EMPLOYEE ☐ CONTRUCTOR		AM	PM			
☐ EMPLOYEE ☐ CONTRUCTOR		AM	PM			
☐ EMPLOYEE ☐ CONTRUCTOR		AM	PM			
☐ EMPLOYEE ☐ CONTRUCTOR		AM	PM			
☐ EMPLOYEE ☐ CONTRUCTOR		AM	PM			

EQUIPMENT ON SITE	NO. OF UNITE	WORKING YES / NO

HIRED EQUIPMENT	NO. OF UNITE	EQUIPMENT RENTED	FROM	RATE

NAME: _____   SIGNATURE: _____

MO TU WE TH FR SA SU
☐   ☐   ☐   ☐   ☐   ☐   ☐

DATE: ___ / ___ / ___

PROJECT: _____

FOREMAN: _____

WEATHER

F° _____ C° _____   _____ AM _____ PM

HOURS DUE TO BAD WEATHER	ISSUED AND DELAYS

NOTE: _____

COMPLETION DATE	DAYS AHEAD OF SCHEDULE	DAYS BEHIND SCHEDULE

SAFETY AND INCIDENTS

SAFETY ISSUES THAT NEED TO BE ADDRESSED	ACCIDENTS / INCIDENTS / STEPS NEEDED TO RESOLVE

SUMMARY OF THE WORK DONE TODAY

_____
_____
_____
_____
_____
_____

IMPORTANT NOTES

_____
_____
_____
_____
_____

NAME	SIGNATURE

TODAY LABOR

INITIALS	TRADE	START	FINISH	PAID HOURS	OVERTIME	COMPANY
☐ EMPLOYEE ☐ CONTRUCTOR		AM	PM			
☐ EMPLOYEE ☐ CONTRUCTOR		AM	PM			
☐ EMPLOYEE ☐ CONTRUCTOR		AM	PM			
☐ EMPLOYEE ☐ CONTRUCTOR		AM	PM			
☐ EMPLOYEE ☐ CONTRUCTOR		AM	PM			
☐ EMPLOYEE ☐ CONTRUCTOR		AM	PM			
☐ EMPLOYEE ☐ CONTRUCTOR		AM	PM			
☐ EMPLOYEE ☐ CONTRUCTOR		AM	PM			

EQUIPMENT ON SITE	NO. OF UNITE	WORKING YES / NO

HIRED EQUIPMENT	NO. OF UNITE	EQUIPMENT RENTED	FROM	RATE

NAME: _____   SIGNATURE: _____

MO TU WE TH FR SA SU
☐   ☐   ☐   ☐   ☐   ☐   ☐

DATE: /   /

PROJECT:

FOREMAN:

WEATHER

F°   C°   AM   PM

HOURS DUE TO BAD WEATHER	ISSUED AND DELAYS

NOTE:

COMPLETION DATE	DAYS AHEAD OF SCHEDULE	DAYS BEHIND SCHEDULE

SAFETY AND INCIDENTS

SAFETY ISSUES THAT NEED TO BE ADDRESSED	ACCIDENTS / INCIDENTS / STEPS NEEDED TO RESOLVE

SUMMARY OF THE WORK DONE TODAY

IMPORTANT NOTES

NAME	SIGNATURE

TODAY LABOR

INITIALS	TRADE	START	FINISH	PAID HOURS	OVERTIME	COMPANY
☐ EMPLOYEE ☐ CONTRUCTOR		AM	PM			
☐ EMPLOYEE ☐ CONTRUCTOR		AM	PM			
☐ EMPLOYEE ☐ CONTRUCTOR		AM	PM			
☐ EMPLOYEE ☐ CONTRUCTOR		AM	PM			
☐ EMPLOYEE ☐ CONTRUCTOR		AM	PM			
☐ EMPLOYEE ☐ CONTRUCTOR		AM	PM			
☐ EMPLOYEE ☐ CONTRUCTOR		AM	PM			
☐ EMPLOYEE ☐ CONTRUCTOR		AM	PM			

EQUIPMENT ON SITE	NO. OF UNITE	WORKING YES / NO

HIRED EQUIPMENT	NO. OF UNITE	EQUIPMENT RENTED	FROM	RATE

NAME: _____   SIGNATURE: _____

MO TU WE TH FR SA SU
☐   ☐   ☐   ☐   ☐   ☐   ☐

DATE: /   /

PROJECT:

FOREMAN:

WEATHER

F°_____ C°_____   _____AM _____PM

HOURS DUE TO BAD WEATHER	ISSUED AND DELAYS

NOTE: _____

COMPLETION DATE	DAYS AHEAD OF SCHEDULE	DAYS BEHIND SCHEDULE

SAFETY AND INCIDENTS

SAFETY ISSUES THAT NEED TO BE ADDRESSED	ACCIDENTS / INCIDENTS / STEPS NEEDED TO RESOLVE

SUMMARY OF THE WORK DONE TODAY

IMPORTANT NOTES

NAME	SIGNATURE

TODAY LABOR

INITIALS	TRADE	START	FINISH	PAID HOURS	OVERTIME	COMPANY
☐ EMPLOYEE ☐ CONTRUCTOR		AM	PM			
☐ EMPLOYEE ☐ CONTRUCTOR		AM	PM			
☐ EMPLOYEE ☐ CONTRUCTOR		AM	PM			
☐ EMPLOYEE ☐ CONTRUCTOR		AM	PM			
☐ EMPLOYEE ☐ CONTRUCTOR		AM	PM			
☐ EMPLOYEE ☐ CONTRUCTOR		AM	PM			
☐ EMPLOYEE ☐ CONTRUCTOR		AM	PM			
☐ EMPLOYEE ☐ CONTRUCTOR		AM	PM			

EQUIPMENT ON SITE	NO. OF UNITE	WORKING YES / NO

HIRED EQUIPMENT	NO. OF UNITE	EQUIPMENT RENTED	FROM	RATE

NAME: _____   SIGNATURE: _____

MO   TU   WE   TH   FR   SA   SU
☐   ☐   ☐   ☐   ☐   ☐   ☐

DATE: /   /

PROJECT:

FOREMAN:

WEATHER   F°____ C°____   ____AM ____PM

HOURS DUE TO BAD WEATHER	ISSUED AND DELAYS

NOTE: _____

COMPLETION DATE	DAYS AHEAD OF SCHEDULE	DAYS BEHIND SCHEDULE

SAFETY AND INCIDENTS

SAFETY ISSUES THAT NEED TO BE ADDRESSED	ACCIDENTS / INCIDENTS / STEPS NEEDED TO RESOLVE

SUMMARY OF THE WORK DONE TODAY

IMPORTANT NOTES

NAME	SIGNATURE

TODAY LABOR

INITIALS	TRADE	START	FINISH	PAID HOURS	OVERTIME	COMPANY
☐ EMPLOYEE ☐ CONTRUCTOR		AM	PM			
☐ EMPLOYEE ☐ CONTRUCTOR		AM	PM			
☐ EMPLOYEE ☐ CONTRUCTOR		AM	PM			
☐ EMPLOYEE ☐ CONTRUCTOR		AM	PM			
☐ EMPLOYEE ☐ CONTRUCTOR		AM	PM			
☐ EMPLOYEE ☐ CONTRUCTOR		AM	PM			
☐ EMPLOYEE ☐ CONTRUCTOR		AM	PM			
☐ EMPLOYEE ☐ CONTRUCTOR		AM	PM			

EQUIPMENT ON SITE	NO. OF UNITE	WORKING YES / NO

HIRED EQUIPMENT	NO. OF UNITE	EQUIPMENT RENTED	FROM	RATE

NAME: _____   SIGNATURE: _____

MO TU WE TH FR SA SU
☐   ☐   ☐   ☐   ☐   ☐   ☐

DATE: /   /

PROJECT:

FOREMAN:

WEATHER

F°_____ C°_____   _____AM _____PM

HOURS DUE TO
BAD WEATHER

ISSUED AND DELAYS

NOTE: _____

COMPLETION DATE	DAYS AHEAD OF SCHEDULE	DAYS BEHIND SCHEDULE

SAFETY AND INCIDENTS

SAFETY ISSUES THAT NEED TO BE ADDRESSED	ACCIDENTS / INCIDENTS / STEPS NEEDED TO RESOLVE

SUMMARY OF THE WORK DONE TODAY

IMPORTANT NOTES

NAME	SIGNATURE

TODAY LABOR

INITIALS	TRADE	START	FINISH	PAID HOURS	OVERTIME	COMPANY
☐ EMPLOYEE ☐ CONTRUCTOR		AM	PM			
☐ EMPLOYEE ☐ CONTRUCTOR		AM	PM			
☐ EMPLOYEE ☐ CONTRUCTOR		AM	PM			
☐ EMPLOYEE ☐ CONTRUCTOR		AM	PM			
☐ EMPLOYEE ☐ CONTRUCTOR		AM	PM			
☐ EMPLOYEE ☐ CONTRUCTOR		AM	PM			
☐ EMPLOYEE ☐ CONTRUCTOR		AM	PM			
☐ EMPLOYEE ☐ CONTRUCTOR		AM	PM			

EQUIPMENT ON SITE	NO. OF UNITE	WORKING YES / NO

HIRED EQUIPMENT	NO. OF UNITE	EQUIPMENT RENTED	FROM	RATE

NAME: _____   SIGNATURE: _____

MO TU WE TH FR SA SU
☐   ☐   ☐   ☐   ☐   ☐   ☐

DATE: /   /

PROJECT:

FOREMAN:

WEATHER

F°_____ C°_____   _____AM _____PM

HOURS DUE TO BAD WEATHER	ISSUED AND DELAYS

NOTE: _____

COMPLETION DATE	DAYS AHEAD OF SCHEDULE	DAYS BEHIND SCHEDULE

SAFETY AND INCIDENTS

SAFETY ISSUES THAT NEED TO BE ADDRESSED	ACCIDENTS / INCIDENTS / STEPS NEEDED TO RESOLVE

SUMMARY OF THE WORK DONE TODAY

IMPORTANT NOTES

NAME	SIGNATURE

TODAY LABOR

INITIALS	TRADE	START	FINISH	PAID HOURS	OVERTIME	COMPANY
☐ EMPLOYEE ☐ CONTRUCTOR		AM	PM			
☐ EMPLOYEE ☐ CONTRUCTOR		AM	PM			
☐ EMPLOYEE ☐ CONTRUCTOR		AM	PM			
☐ EMPLOYEE ☐ CONTRUCTOR		AM	PM			
☐ EMPLOYEE ☐ CONTRUCTOR		AM	PM			
☐ EMPLOYEE ☐ CONTRUCTOR		AM	PM			
☐ EMPLOYEE ☐ CONTRUCTOR		AM	PM			
☐ EMPLOYEE ☐ CONTRUCTOR		AM	PM			

EQUIPMENT ON SITE	NO. OF UNITE	WORKING YES / NO

HIRED EQUIPMENT	NO. OF UNITE	EQUIPMENT RENTED	FROM	RATE

NAME: _____   SIGNATURE: _____

MO TU WE TH FR SA SU
☐   ☐   ☐   ☐   ☐   ☐   ☐

DATE: ___ / ___ / ___

PROJECT: _____

FOREMAN: _____

WEATHER

F° _____ C° _____   _____ AM _____ PM

HOURS DUE TO BAD WEATHER

ISSUED AND DELAYS

NOTE: _____

COMPLETION DATE	DAYS AHEAD OF SCHEDULE	DAYS BEHIND SCHEDULE

SAFETY AND INCIDENTS

SAFETY ISSUES THAT NEED TO BE ADDRESSED	ACCIDENTS / INCIDENTS / STEPS NEEDED TO RESOLVE

SUMMARY OF THE WORK DONE TODAY

_____
_____
_____
_____
_____

IMPORTANT NOTES

_____
_____
_____
_____
_____

NAME	SIGNATURE

TODAY LABOR

INITIALS	TRADE	START	FINISH	PAID HOURS	OVERTIME	COMPANY
☐ EMPLOYEE ☐ CONTRUCTOR		AM	PM			
☐ EMPLOYEE ☐ CONTRUCTOR		AM	PM			
☐ EMPLOYEE ☐ CONTRUCTOR		AM	PM			
☐ EMPLOYEE ☐ CONTRUCTOR		AM	PM			
☐ EMPLOYEE ☐ CONTRUCTOR		AM	PM			
☐ EMPLOYEE ☐ CONTRUCTOR		AM	PM			
☐ EMPLOYEE ☐ CONTRUCTOR		AM	PM			
☐ EMPLOYEE ☐ CONTRUCTOR		AM	PM			

EQUIPMENT ON SITE	NO. OF UNITE	WORKING YES / NO

HIRED EQUIPMENT	NO. OF UNITE	EQUIPMENT RENTED	FROM	RATE

NAME: _____   SIGNATURE: _____

MO TU WE TH FR SA SU
☐   ☐   ☐   ☐   ☐   ☐   ☐   DATE: /   /

PROJECT: FOREMAN:

WEATHER   HOURS DUE TO   ISSUED AND DELAYS
                                        BAD WEATHER

           F°   C°   ____AM   ____PM

NOTE: _____

COMPLETION DATE	DAYS AHEAD OF SCHEDULE	DAYS BEHIND SCHEDULE

SAFETY AND INCIDENTS

SAFETY ISSUES THAT NEED TO BE ADDRESSED	ACCIDENTS / INCIDENTS / STEPS NEEDED TO RESOLVE

SUMMARY OF THE WORK DONE TODAY

IMPORTANT NOTES

NAME	SIGNATURE

TODAY LABOR

INITIALS	TRADE	START	FINISH	PAID HOURS	OVERTIME	COMPANY
☐ EMPLOYEE ☐ CONTRUCTOR		AM	PM			
☐ EMPLOYEE ☐ CONTRUCTOR		AM	PM			
☐ EMPLOYEE ☐ CONTRUCTOR		AM	PM			
☐ EMPLOYEE ☐ CONTRUCTOR		AM	PM			
☐ EMPLOYEE ☐ CONTRUCTOR		AM	PM			
☐ EMPLOYEE ☐ CONTRUCTOR		AM	PM			
☐ EMPLOYEE ☐ CONTRUCTOR		AM	PM			
☐ EMPLOYEE ☐ CONTRUCTOR		AM	PM			

EQUIPMENT ON SITE	NO. OF UNITE	WORKING YES / NO

HIRED EQUIPMENT	NO. OF UNITE	EQUIPMENT RENTED	FROM	RATE

NAME: _____   SIGNATURE: _____

MO TU WE TH FR SA SU
☐   ☐   ☐   ☐   ☐   ☐   ☐

DATE: __ / __ / __

PROJECT: _____

FOREMAN: _____

WEATHER

F° ____ C° ____ ____ AM ____ PM

HOURS DUE TO BAD WEATHER	ISSUED AND DELAYS
	_____

NOTE: _____

COMPLETION DATE	DAYS AHEAD OF SCHEDULE	DAYS BEHIND SCHEDULE

SAFETY AND INCIDENTS

SAFETY ISSUES THAT NEED TO BE ADDRESSED	ACCIDENTS / INCIDENTS / STEPS NEEDED TO RESOLVE
_____	_____
_____	_____
_____	_____
_____	_____

SUMMARY OF THE WORK DONE TODAY

_____
_____
_____
_____
_____

IMPORTANT NOTES

_____
_____
_____
_____
_____

NAME	SIGNATURE

TODAY LABOR

INITIALS	TRADE	START	FINISH	PAID HOURS	OVERTIME	COMPANY
☐ EMPLOYEE ☐ CONTRUCTOR		AM	PM			
☐ EMPLOYEE ☐ CONTRUCTOR		AM	PM			
☐ EMPLOYEE ☐ CONTRUCTOR		AM	PM			
☐ EMPLOYEE ☐ CONTRUCTOR		AM	PM			
☐ EMPLOYEE ☐ CONTRUCTOR		AM	PM			
☐ EMPLOYEE ☐ CONTRUCTOR		AM	PM			
☐ EMPLOYEE ☐ CONTRUCTOR		AM	PM			
☐ EMPLOYEE ☐ CONTRUCTOR		AM	PM			

EQUIPMENT ON SITE	NO. OF UNITE	WORKING YES / NO

HIRED EQUIPMENT	NO. OF UNITE	EQUIPMENT RENTED	FROM	RATE

NAME: _____   SIGNATURE: _____

MO TU WE TH FR SA SU
☐   ☐   ☐   ☐   ☐   ☐   ☐

DATE: /   /

PROJECT:

FOREMAN:

WEATHER

F°____ C°____ ____AM ____PM

HOURS DUE TO BAD WEATHER	ISSUED AND DELAYS

NOTE: _____

COMPLETION DATE	DAYS AHEAD OF SCHEDULE	DAYS BEHIND SCHEDULE

SAFETY AND INCIDENTS

SAFETY ISSUES THAT NEED TO BE ADDRESSED	ACCIDENTS / INCIDENTS / STEPS NEEDED TO RESOLVE

SUMMARY OF THE WORK DONE TODAY

IMPORTANT NOTES

NAME	SIGNATURE

TODAY LABOR

INITIALS	TRADE	START	FINISH	PAID HOURS	OVERTIME	COMPANY
☐ EMPLOYEE ☐ CONTRUCTOR		AM	PM			
☐ EMPLOYEE ☐ CONTRUCTOR		AM	PM			
☐ EMPLOYEE ☐ CONTRUCTOR		AM	PM			
☐ EMPLOYEE ☐ CONTRUCTOR		AM	PM			
☐ EMPLOYEE ☐ CONTRUCTOR		AM	PM			
☐ EMPLOYEE ☐ CONTRUCTOR		AM	PM			
☐ EMPLOYEE ☐ CONTRUCTOR		AM	PM			
☐ EMPLOYEE ☐ CONTRUCTOR		AM	PM			

EQUIPMENT ON SITE	NO. OF UNITE	WORKING YES / NO

HIRED EQUIPMENT	NO. OF UNITE	EQUIPMENT RENTED	FROM	RATE

NAME: _____   SIGNATURE: _____

MO   TU   WE   TH   FR   SA   SU
☐   ☐   ☐   ☐   ☐   ☐   ☐

DATE: /   /

PROJECT:

FOREMAN:

WEATHER   HOURS DUE TO   ISSUED AND DELAYS
                                     BAD WEATHER

        F°_____ C°_____ _____AM _____PM

NOTE: _____

COMPLETION DATE	DAYS AHEAD OF SCHEDULE	DAYS BEHIND SCHEDULE

SAFETY AND INCIDENTS

SAFETY ISSUES THAT NEED TO BE ADDRESSED	ACCIDENTS / INCIDENTS / STEPS NEEDED TO RESOLVE
_____	_____
_____	_____
_____	_____
_____	_____

SUMMARY OF THE WORK DONE TODAY

_____
_____
_____
_____
_____

IMPORTANT NOTES

_____
_____
_____
_____
_____

NAME	SIGNATURE

TODAY LABOR

INITIALS	TRADE	START	FINISH	PAID HOURS	OVERTIME	COMPANY
☐ EMPLOYEE ☐ CONTRUCTOR		AM	PM			
☐ EMPLOYEE ☐ CONTRUCTOR		AM	PM			
☐ EMPLOYEE ☐ CONTRUCTOR		AM	PM			
☐ EMPLOYEE ☐ CONTRUCTOR		AM	PM			
☐ EMPLOYEE ☐ CONTRUCTOR		AM	PM			
☐ EMPLOYEE ☐ CONTRUCTOR		AM	PM			
☐ EMPLOYEE ☐ CONTRUCTOR		AM	PM			
☐ EMPLOYEE ☐ CONTRUCTOR		AM	PM			

EQUIPMENT ON SITE	NO. OF UNITE	WORKING YES / NO

HIRED EQUIPMENT	NO. OF UNITE	EQUIPMENT RENTED	FROM	RATE

NAME: _____   SIGNATURE: _____

MO   TU   WE   TH   FR   SA   SU
☐   ☐   ☐   ☐   ☐   ☐   ☐

DATE: /   /

PROJECT: FOREMAN:

WEATHER   ☁🌧 ⛅ ☁ 🌨 ☀ 🌦 ⛈

F°____   C°____   ____AM   ____PM

HOURS DUE TO BAD WEATHER	ISSUED AND DELAYS

NOTE: _____

COMPLETION DATE	DAYS AHEAD OF SCHEDULE	DAYS BEHIND SCHEDULE

SAFETY AND INCIDENTS

SAFETY ISSUES THAT NEED TO BE ADDRESSED	ACCIDENTS / INCIDENTS / STEPS NEEDED TO RESOLVE

SUMMARY OF THE WORK DONE TODAY

IMPORTANT NOTES

NAME	SIGNATURE

TODAY LABOR

INITIALS	TRADE	START	FINISH	PAID HOURS	OVERTIME	COMPANY
☐ EMPLOYEE ☐ CONTRUCTOR		AM	PM			
☐ EMPLOYEE ☐ CONTRUCTOR		AM	PM			
☐ EMPLOYEE ☐ CONTRUCTOR		AM	PM			
☐ EMPLOYEE ☐ CONTRUCTOR		AM	PM			
☐ EMPLOYEE ☐ CONTRUCTOR		AM	PM			
☐ EMPLOYEE ☐ CONTRUCTOR		AM	PM			
☐ EMPLOYEE ☐ CONTRUCTOR		AM	PM			
☐ EMPLOYEE ☐ CONTRUCTOR		AM	PM			

EQUIPMENT ON SITE	NO. OF UNITE	WORKING YES / NO

HIRED EQUIPMENT	NO. OF UNITE	EQUIPMENT RENTED	FROM	RATE

NAME: _____   SIGNATURE: _____

MO TU WE TH FR SA SU
☐   ☐   ☐   ☐   ☐   ☐   ☐

DATE: /   /

PROJECT: FOREMAN:

WEATHER   F°____ C°____ ____AM ____PM

HOURS DUE TO BAD WEATHER

ISSUED AND DELAYS

NOTE:

COMPLETION DATE	DAYS AHEAD OF SCHEDULE	DAYS BEHIND SCHEDULE

SAFETY AND INCIDENTS

SAFETY ISSUES THAT NEED TO BE ADDRESSED	ACCIDENTS / INCIDENTS / STEPS NEEDED TO RESOLVE

SUMMARY OF THE WORK DONE TODAY

IMPORTANT NOTES

NAME	SIGNATURE

TODAY LABOR

INITIALS	TRADE	START	FINISH	PAID HOURS	OVERTIME	COMPANY
☐ EMPLOYEE ☐ CONTRUCTOR		AM	PM			
☐ EMPLOYEE ☐ CONTRUCTOR		AM	PM			
☐ EMPLOYEE ☐ CONTRUCTOR		AM	PM			
☐ EMPLOYEE ☐ CONTRUCTOR		AM	PM			
☐ EMPLOYEE ☐ CONTRUCTOR		AM	PM			
☐ EMPLOYEE ☐ CONTRUCTOR		AM	PM			
☐ EMPLOYEE ☐ CONTRUCTOR		AM	PM			
☐ EMPLOYEE ☐ CONTRUCTOR		AM	PM			

EQUIPMENT ON SITE	NO. OF UNITE	WORKING YES / NO

HIRED EQUIPMENT	NO. OF UNITE	EQUIPMENT RENTED	FROM	RATE

NAME: _____   SIGNATURE: _____

MO TU WE TH FR SA SU
☐   ☐   ☐   ☐   ☐   ☐   ☐

DATE: /   /

PROJECT:

FOREMAN:

WEATHER

F°____ C°____   ____AM   ____PM

HOURS DUE TO BAD WEATHER	ISSUED AND DELAYS

NOTE: _____

COMPLETION DATE	DAYS AHEAD OF SCHEDULE	DAYS BEHIND SCHEDULE

SAFETY AND INCIDENTS

SAFETY ISSUES THAT NEED TO BE ADDRESSED	ACCIDENTS / INCIDENTS / STEPS NEEDED TO RESOLVE

SUMMARY OF THE WORK DONE TODAY

IMPORTANT NOTES

NAME	SIGNATURE

TODAY LABOR

INITIALS	TRADE	START	FINISH	PAID HOURS	OVERTIME	COMPANY
☐ EMPLOYEE ☐ CONTRUCTOR		AM	PM			
☐ EMPLOYEE ☐ CONTRUCTOR		AM	PM			
☐ EMPLOYEE ☐ CONTRUCTOR		AM	PM			
☐ EMPLOYEE ☐ CONTRUCTOR		AM	PM			
☐ EMPLOYEE ☐ CONTRUCTOR		AM	PM			
☐ EMPLOYEE ☐ CONTRUCTOR		AM	PM			
☐ EMPLOYEE ☐ CONTRUCTOR		AM	PM			
☐ EMPLOYEE ☐ CONTRUCTOR		AM	PM			

EQUIPMENT ON SITE	NO. OF UNITE	WORKING YES / NO

HIRED EQUIPMENT	NO. OF UNITE	EQUIPMENT RENTED	FROM	RATE

NAME: _____   SIGNATURE: _____

MO TU WE TH FR SA SU
☐   ☐   ☐   ☐   ☐   ☐   ☐

DATE: __ / __ / __

PROJECT:

FOREMAN:

WEATHER

F° ____ C° ____   ____ AM ____ PM

| HOURS DUE TO BAD WEATHER | ISSUED AND DELAYS |

NOTE: _____

COMPLETION DATE	DAYS AHEAD OF SCHEDULE	DAYS BEHIND SCHEDULE

SAFETY AND INCIDENTS

SAFETY ISSUES THAT NEED TO BE ADDRESSED	ACCIDENTS / INCIDENTS / STEPS NEEDED TO RESOLVE

SUMMARY OF THE WORK DONE TODAY

IMPORTANT NOTES

NAME	SIGNATURE

TODAY LABOR

INITIALS	TRADE	START	FINISH	PAID HOURS	OVERTIME	COMPANY
☐ EMPLOYEE ☐ CONTRUCTOR		AM	PM			
☐ EMPLOYEE ☐ CONTRUCTOR		AM	PM			
☐ EMPLOYEE ☐ CONTRUCTOR		AM	PM			
☐ EMPLOYEE ☐ CONTRUCTOR		AM	PM			
☐ EMPLOYEE ☐ CONTRUCTOR		AM	PM			
☐ EMPLOYEE ☐ CONTRUCTOR		AM	PM			
☐ EMPLOYEE ☐ CONTRUCTOR		AM	PM			
☐ EMPLOYEE ☐ CONTRUCTOR		AM	PM			

EQUIPMENT ON SITE	NO. OF UNITE	WORKING YES / NO

HIRED EQUIPMENT	NO. OF UNITE	EQUIPMENT RENTED	FROM	RATE

NAME: _____   SIGNATURE: _____

MO TU WE TH FR SA SU
☐   ☐   ☐   ☐   ☐   ☐   ☐

DATE: /   /

PROJECT:

FOREMAN:

WEATHER

F°_____   C°_____   _____AM   _____PM

HOURS DUE TO BAD WEATHER	ISSUED AND DELAYS

NOTE: _____

COMPLETION DATE	DAYS AHEAD OF SCHEDULE	DAYS BEHIND SCHEDULE

SAFETY AND INCIDENTS

SAFETY ISSUES THAT NEED TO BE ADDRESSED	ACCIDENTS / INCIDENTS / STEPS NEEDED TO RESOLVE

SUMMARY OF THE WORK DONE TODAY

IMPORTANT NOTES

NAME	SIGNATURE

TODAY LABOR

INITIALS	TRADE	START	FINISH	PAID HOURS	OVERTIME	COMPANY
☐ EMPLOYEE ☐ CONTRUCTOR		AM	PM			
☐ EMPLOYEE ☐ CONTRUCTOR		AM	PM			
☐ EMPLOYEE ☐ CONTRUCTOR		AM	PM			
☐ EMPLOYEE ☐ CONTRUCTOR		AM	PM			
☐ EMPLOYEE ☐ CONTRUCTOR		AM	PM			
☐ EMPLOYEE ☐ CONTRUCTOR		AM	PM			
☐ EMPLOYEE ☐ CONTRUCTOR		AM	PM			
☐ EMPLOYEE ☐ CONTRUCTOR		AM	PM			

EQUIPMENT ON SITE	NO. OF UNITE	WORKING YES / NO

HIRED EQUIPMENT	NO. OF UNITE	EQUIPMENT RENTED	FROM	RATE

NAME: _____   SIGNATURE: _____

MO TU WE TH FR SA SU
☐   ☐   ☐   ☐   ☐   ☐   ☐

DATE: ___/___/___

PROJECT:

FOREMAN:

WEATHER

F°_____ C°_____ _____AM _____PM

HOURS DUE TO BAD WEATHER	ISSUED AND DELAYS

NOTE: _____

COMPLETION DATE	DAYS AHEAD OF SCHEDULE	DAYS BEHIND SCHEDULE

SAFETY AND INCIDENTS

SAFETY ISSUES THAT NEED TO BE ADDRESSED	ACCIDENTS / INCIDENTS / STEPS NEEDED TO RESOLVE

SUMMARY OF THE WORK DONE TODAY

IMPORTANT NOTES

NAME	SIGNATURE

TODAY LABOR

INITIALS	TRADE	START	FINISH	PAID HOURS	OVERTIME	COMPANY
☐ EMPLOYEE ☐ CONTRUCTOR		AM	PM			
☐ EMPLOYEE ☐ CONTRUCTOR		AM	PM			
☐ EMPLOYEE ☐ CONTRUCTOR		AM	PM			
☐ EMPLOYEE ☐ CONTRUCTOR		AM	PM			
☐ EMPLOYEE ☐ CONTRUCTOR		AM	PM			
☐ EMPLOYEE ☐ CONTRUCTOR		AM	PM			
☐ EMPLOYEE ☐ CONTRUCTOR		AM	PM			
☐ EMPLOYEE ☐ CONTRUCTOR		AM	PM			

EQUIPMENT ON SITE	NO. OF UNITE	WORKING YES / NO

HIRED EQUIPMENT	NO. OF UNITE	EQUIPMENT RENTED	FROM	RATE

NAME: _____ SIGNATURE: _____

MO TU WE TH FR SA SU
☐   ☐   ☐   ☐   ☐   ☐   ☐

DATE: /   /

PROJECT:

FOREMAN:

WEATHER

F° _____ C° _____   _____ AM _____ PM

HOURS DUE TO
BAD WEATHER

ISSUED AND DELAYS

NOTE: _____

COMPLETION DATE	DAYS AHEAD OF SCHEDULE	DAYS BEHIND SCHEDULE

SAFETY AND INCIDENTS

SAFETY ISSUES THAT NEED TO BE ADDRESSED	ACCIDENTS / INCIDENTS / STEPS NEEDED TO RESOLVE

SUMMARY OF THE WORK DONE TODAY

IMPORTANT NOTES

NAME	SIGNATURE

TODAY LABOR

INITIALS	TRADE	START	FINISH	PAID HOURS	OVERTIME	COMPANY
☐ EMPLOYEE ☐ CONTRUCTOR		AM	PM			
☐ EMPLOYEE ☐ CONTRUCTOR		AM	PM			
☐ EMPLOYEE ☐ CONTRUCTOR		AM	PM			
☐ EMPLOYEE ☐ CONTRUCTOR		AM	PM			
☐ EMPLOYEE ☐ CONTRUCTOR		AM	PM			
☐ EMPLOYEE ☐ CONTRUCTOR		AM	PM			
☐ EMPLOYEE ☐ CONTRUCTOR		AM	PM			
☐ EMPLOYEE ☐ CONTRUCTOR		AM	PM			

EQUIPMENT ON SITE	NO. OF UNITE	WORKING YES / NO

HIRED EQUIPMENT	NO. OF UNITE	EQUIPMENT RENTED	FROM	RATE

NAME: _____   SIGNATURE: _____

MO TU WE TH FR SA SU
☐   ☐   ☐   ☐   ☐   ☐   ☐

DATE: /   /

PROJECT: FOREMAN:

WEATHER

F°____ C°____   ____AM ____PM

| HOURS DUE TO BAD WEATHER | ISSUED AND DELAYS |

NOTE: _____

COMPLETION DATE	DAYS AHEAD OF SCHEDULE	DAYS BEHIND SCHEDULE

SAFETY AND INCIDENTS

SAFETY ISSUES THAT NEED TO BE ADDRESSED	ACCIDENTS / INCIDENTS / STEPS NEEDED TO RESOLVE

SUMMARY OF THE WORK DONE TODAY

_____
_____
_____
_____
_____

IMPORTANT NOTES

_____
_____
_____
_____
_____

NAME	SIGNATURE

TODAY LABOR

INITIALS	TRADE	START	FINISH	PAID HOURS	OVERTIME	COMPANY
☐ EMPLOYEE ☐ CONTRUCTOR		AM	PM			
☐ EMPLOYEE ☐ CONTRUCTOR		AM	PM			
☐ EMPLOYEE ☐ CONTRUCTOR		AM	PM			
☐ EMPLOYEE ☐ CONTRUCTOR		AM	PM			
☐ EMPLOYEE ☐ CONTRUCTOR		AM	PM			
☐ EMPLOYEE ☐ CONTRUCTOR		AM	PM			
☐ EMPLOYEE ☐ CONTRUCTOR		AM	PM			
☐ EMPLOYEE ☐ CONTRUCTOR		AM	PM			

EQUIPMENT ON SITE	NO. OF UNITE	WORKING YES / NO

HIRED EQUIPMENT	NO. OF UNITE	EQUIPMENT RENTED	FROM	RATE

NAME: _____   SIGNATURE: _____

MO   TU   WE   TH   FR   SA   SU
☐   ☐   ☐   ☐   ☐   ☐   ☐

DATE: /   /

PROJECT: FOREMAN:

WEATHER ☁ ⛅ ☁ 🌨 ☀ 🌧 ⛈

F° ____ C° ____ ____ AM ____ PM

| HOURS DUE TO BAD WEATHER | ISSUED AND DELAYS |

NOTE: _____

COMPLETION DATE	DAYS AHEAD OF SCHEDULE	DAYS BEHIND SCHEDULE

SAFETY AND INCIDENTS

SAFETY ISSUES THAT NEED TO BE ADDRESSED	ACCIDENTS / INCIDENTS / STEPS NEEDED TO RESOLVE

SUMMARY OF THE WORK DONE TODAY

IMPORTANT NOTES

NAME	SIGNATURE

TODAY LABOR

INITIALS	TRADE	START	FINISH	PAID HOURS	OVERTIME	COMPANY
☐ EMPLOYEE ☐ CONTRUCTOR		AM	PM			
☐ EMPLOYEE ☐ CONTRUCTOR		AM	PM			
☐ EMPLOYEE ☐ CONTRUCTOR		AM	PM			
☐ EMPLOYEE ☐ CONTRUCTOR		AM	PM			
☐ EMPLOYEE ☐ CONTRUCTOR		AM	PM			
☐ EMPLOYEE ☐ CONTRUCTOR		AM	PM			
☐ EMPLOYEE ☐ CONTRUCTOR		AM	PM			
☐ EMPLOYEE ☐ CONTRUCTOR		AM	PM			

EQUIPMENT ON SITE	NO. OF UNITE	WORKING YES / NO

HIRED EQUIPMENT	NO. OF UNITE	EQUIPMENT RENTED	FROM	RATE

NAME: _____   SIGNATURE: _____

MO TU WE TH FR SA SU
☐ ☐ ☐ ☐ ☐ ☐ ☐

DATE: / /

PROJECT:

FOREMAN:

WEATHER

F°_____ C°_____ _____AM _____PM

HOURS DUE TO BAD WEATHER

ISSUED AND DELAYS

NOTE: _____

COMPLETION DATE	DAYS AHEAD OF SCHEDULE	DAYS BEHIND SCHEDULE

SAFETY AND INCIDENTS

SAFETY ISSUES THAT NEED TO BE ADDRESSED	ACCIDENTS / INCIDENTS / STEPS NEEDED TO RESOLVE

SUMMARY OF THE WORK DONE TODAY

IMPORTANT NOTES

NAME	SIGNATURE

TODAY LABOR

INITIALS	TRADE	START	FINISH	PAID HOURS	OVERTIME	COMPANY
☐ EMPLOYEE ☐ CONTRUCTOR		AM	PM			
☐ EMPLOYEE ☐ CONTRUCTOR		AM	PM			
☐ EMPLOYEE ☐ CONTRUCTOR		AM	PM			
☐ EMPLOYEE ☐ CONTRUCTOR		AM	PM			
☐ EMPLOYEE ☐ CONTRUCTOR		AM	PM			
☐ EMPLOYEE ☐ CONTRUCTOR		AM	PM			
☐ EMPLOYEE ☐ CONTRUCTOR		AM	PM			
☐ EMPLOYEE ☐ CONTRUCTOR		AM	PM			

EQUIPMENT ON SITE	NO. OF UNITE	WORKING YES / NO

HIRED EQUIPMENT	NO. OF UNITE	EQUIPMENT RENTED	FROM	RATE

NAME: _____   SIGNATURE: _____

| MO | TU | WE | TH | FR | SA | SU | | DATE: / / |
| □ | □ | □ | □ | □ | □ | □ | | |

| PROJECT: | FOREMAN: |

WEATHER F°_____ C°_____ _____AM _____PM

| | HOURS DUE TO BAD WEATHER | ISSUED AND DELAYS |

NOTE: _____

| COMPLETION DATE | DAYS AHEAD OF SCHEDULE | DAYS BEHIND SCHEDULE |
| | | |

SAFETY AND INCIDENTS

| SAFETY ISSUES THAT NEED TO BE ADDRESSED | ACCIDENTS / INCIDENTS / STEPS NEEDED TO RESOLVE |
| | |

SUMMARY OF THE WORK DONE TODAY

IMPORTANT NOTES

| NAME | SIGNATURE |
| | |

TODAY LABOR

INITIALS	TRADE	START	FINISH	PAID HOURS	OVERTIME	COMPANY
☐ EMPLOYEE ☐ CONTRUCTOR		AM	PM			
☐ EMPLOYEE ☐ CONTRUCTOR		AM	PM			
☐ EMPLOYEE ☐ CONTRUCTOR		AM	PM			
☐ EMPLOYEE ☐ CONTRUCTOR		AM	PM			
☐ EMPLOYEE ☐ CONTRUCTOR		AM	PM			
☐ EMPLOYEE ☐ CONTRUCTOR		AM	PM			
☐ EMPLOYEE ☐ CONTRUCTOR		AM	PM			
☐ EMPLOYEE ☐ CONTRUCTOR		AM	PM			

EQUIPMENT ON SITE	NO. OF UNITE	WORKING YES / NO

HIRED EQUIPMENT	NO. OF UNITE	EQUIPMENT RENTED	FROM	RATE

NAME: _____   SIGNATURE: _____

MO TU WE TH FR SA SU
☐   ☐   ☐   ☐   ☐   ☐   ☐

DATE: ___ / ___ / ___

PROJECT:

FOREMAN:

WEATHER

F° ____ C° ____ ____ AM ____ PM

HOURS DUE TO BAD WEATHER	ISSUED AND DELAYS

NOTE: _____

COMPLETION DATE	DAYS AHEAD OF SCHEDULE	DAYS BEHIND SCHEDULE

SAFETY AND INCIDENTS

SAFETY ISSUES THAT NEED TO BE ADDRESSED	ACCIDENTS / INCIDENTS / STEPS NEEDED TO RESOLVE

SUMMARY OF THE WORK DONE TODAY

IMPORTANT NOTES

NAME	SIGNATURE

TODAY LABOR

INITIALS	TRADE	START	FINISH	PAID HOURS	OVERTIME	COMPANY
☐ EMPLOYEE ☐ CONTRUCTOR		AM	PM			
☐ EMPLOYEE ☐ CONTRUCTOR		AM	PM			
☐ EMPLOYEE ☐ CONTRUCTOR		AM	PM			
☐ EMPLOYEE ☐ CONTRUCTOR		AM	PM			
☐ EMPLOYEE ☐ CONTRUCTOR		AM	PM			
☐ EMPLOYEE ☐ CONTRUCTOR		AM	PM			
☐ EMPLOYEE ☐ CONTRUCTOR		AM	PM			
☐ EMPLOYEE ☐ CONTRUCTOR		AM	PM			

EQUIPMENT ON SITE	NO. OF UNITE	WORKING YES / NO

HIRED EQUIPMENT	NO. OF UNITE	EQUIPMENT RENTED	FROM	RATE

NAME: _____   SIGNATURE: _____

MO TU WE TH FR SA SU
☐   ☐   ☐   ☐   ☐   ☐   ☐

DATE: /   /

PROJECT:

FOREMAN:

WEATHER

F°____ C°____   ____AM ____PM

| HOURS DUE TO BAD WEATHER | ISSUED AND DELAYS |

NOTE: _____

COMPLETION DATE	DAYS AHEAD OF SCHEDULE	DAYS BEHIND SCHEDULE

SAFETY AND INCIDENTS

SAFETY ISSUES THAT NEED TO BE ADDRESSED	ACCIDENTS / INCIDENTS / STEPS NEEDED TO RESOLVE

SUMMARY OF THE WORK DONE TODAY

IMPORTANT NOTES

NAME	SIGNATURE

TODAY LABOR

INITIALS	TRADE	START	FINISH	PAID HOURS	OVERTIME	COMPANY
☐ EMPLOYEE ☐ CONTRUCTOR		AM	PM			
☐ EMPLOYEE ☐ CONTRUCTOR		AM	PM			
☐ EMPLOYEE ☐ CONTRUCTOR		AM	PM			
☐ EMPLOYEE ☐ CONTRUCTOR		AM	PM			
☐ EMPLOYEE ☐ CONTRUCTOR		AM	PM			
☐ EMPLOYEE ☐ CONTRUCTOR		AM	PM			
☐ EMPLOYEE ☐ CONTRUCTOR		AM	PM			
☐ EMPLOYEE ☐ CONTRUCTOR		AM	PM			

EQUIPMENT ON SITE	NO. OF UNITE	WORKING YES / NO

HIRED EQUIPMENT	NO. OF UNITE	EQUIPMENT RENTED	FROM	RATE

NAME: _____   SIGNATURE: _____

MO   TU   WE   TH   FR   SA   SU
☐   ☐   ☐   ☐   ☐   ☐   ☐

DATE: /   /

PROJECT:

FOREMAN:

WEATHER

F°_____ C°_____ _____AM _____PM

HOURS DUE TO
BAD WEATHER

ISSUED AND DELAYS

NOTE: _____

COMPLETION DATE	DAYS AHEAD OF SCHEDULE	DAYS BEHIND SCHEDULE

SAFETY AND INCIDENTS

SAFETY ISSUES THAT NEED TO BE ADDRESSED	ACCIDENTS / INCIDENTS / STEPS NEEDED TO RESOLVE

SUMMARY OF THE WORK DONE TODAY

IMPORTANT NOTES

NAME	SIGNATURE

TODAY LABOR

INITIALS	TRADE	START	FINISH	PAID HOURS	OVERTIME	COMPANY
☐ EMPLOYEE ☐ CONTRUCTOR		AM	PM			
☐ EMPLOYEE ☐ CONTRUCTOR		AM	PM			
☐ EMPLOYEE ☐ CONTRUCTOR		AM	PM			
☐ EMPLOYEE ☐ CONTRUCTOR		AM	PM			
☐ EMPLOYEE ☐ CONTRUCTOR		AM	PM			
☐ EMPLOYEE ☐ CONTRUCTOR		AM	PM			
☐ EMPLOYEE ☐ CONTRUCTOR		AM	PM			
☐ EMPLOYEE ☐ CONTRUCTOR		AM	PM			

EQUIPMENT ON SITE	NO. OF UNITE	WORKING YES / NO

HIRED EQUIPMENT	NO. OF UNITE	EQUIPMENT RENTED	FROM	RATE

NAME: _____   SIGNATURE: _____

MO   TU   WE   TH   FR   SA   SU
☐   ☐   ☐   ☐   ☐   ☐   ☐

DATE: _____ / _____ / _____

PROJECT:

FOREMAN:

WEATHER

F° _____   C° _____   _____ AM   _____ PM

HOURS DUE TO BAD WEATHER	ISSUED AND DELAYS

NOTE: _____

COMPLETION DATE	DAYS AHEAD OF SCHEDULE	DAYS BEHIND SCHEDULE

SAFETY AND INCIDENTS

SAFETY ISSUES THAT NEED TO BE ADDRESSED	ACCIDENTS / INCIDENTS / STEPS NEEDED TO RESOLVE

SUMMARY OF THE WORK DONE TODAY

IMPORTANT NOTES

NAME	SIGNATURE

TODAY LABOR

INITIALS	TRADE	START	FINISH	PAID HOURS	OVERTIME	COMPANY
☐ EMPLOYEE ☐ CONTRUCTOR		AM	PM			
☐ EMPLOYEE ☐ CONTRUCTOR		AM	PM			
☐ EMPLOYEE ☐ CONTRUCTOR		AM	PM			
☐ EMPLOYEE ☐ CONTRUCTOR		AM	PM			
☐ EMPLOYEE ☐ CONTRUCTOR		AM	PM			
☐ EMPLOYEE ☐ CONTRUCTOR		AM	PM			
☐ EMPLOYEE ☐ CONTRUCTOR		AM	PM			
☐ EMPLOYEE ☐ CONTRUCTOR		AM	PM			

EQUIPMENT ON SITE	NO. OF UNITE	WORKING YES / NO

HIRED EQUIPMENT	NO. OF UNITE	EQUIPMENT RENTED	FROM	RATE

NAME: _____   SIGNATURE: _____

MO   TU   WE   TH   FR   SA   SU
☐   ☐   ☐   ☐   ☐   ☐   ☐   DATE: /   /

PROJECT: FOREMAN:

WEATHER   HOURS DUE TO   ISSUED AND DELAYS
                                       BAD WEATHER
          F°   C°   ___ AM ___ PM

NOTE: _____

COMPLETION DATE	DAYS AHEAD OF SCHEDULE	DAYS BEHIND SCHEDULE

SAFETY AND INCIDENTS

SAFETY ISSUES THAT NEED TO BE ADDRESSED	ACCIDENTS / INCIDENTS / STEPS NEEDED TO RESOLVE

SUMMARY OF THE WORK DONE TODAY

_____
_____
_____
_____
_____

IMPORTANT NOTES

_____
_____
_____
_____
_____

NAME	SIGNATURE

TODAY LABOR

INITIALS	TRADE	START	FINISH	PAID HOURS	OVERTIME	COMPANY
☐ EMPLOYEE ☐ CONTRUCTOR		AM	PM			
☐ EMPLOYEE ☐ CONTRUCTOR		AM	PM			
☐ EMPLOYEE ☐ CONTRUCTOR		AM	PM			
☐ EMPLOYEE ☐ CONTRUCTOR		AM	PM			
☐ EMPLOYEE ☐ CONTRUCTOR		AM	PM			
☐ EMPLOYEE ☐ CONTRUCTOR		AM	PM			
☐ EMPLOYEE ☐ CONTRUCTOR		AM	PM			
☐ EMPLOYEE ☐ CONTRUCTOR		AM	PM			

EQUIPMENT ON SITE	NO. OF UNITE	WORKING YES / NO

HIRED EQUIPMENT	NO. OF UNITE	EQUIPMENT RENTED	FROM	RATE

NAME: _____   SIGNATURE: _____

MO   TU   WE   TH   FR   SA   SU
☐   ☐   ☐   ☐   ☐   ☐   ☐

DATE: /   /

PROJECT:

FOREMAN:

WEATHER   ☁   ⛅   ☁   🌦   ☀   🌧   ⛈

F°____   C°____   ____AM   ____PM

HOURS DUE TO
BAD WEATHER

ISSUED AND DELAYS

NOTE: _____

COMPLETION DATE	DAYS AHEAD OF SCHEDULE	DAYS BEHIND SCHEDULE

SAFETY AND INCIDENTS

SAFETY ISSUES THAT NEED TO BE ADDRESSED	ACCIDENTS / INCIDENTS / STEPS NEEDED TO RESOLVE

SUMMARY OF THE WORK DONE TODAY

IMPORTANT NOTES

NAME	SIGNATURE

TODAY LABOR

INITIALS	TRADE	START	FINISH	PAID HOURS	OVERTIME	COMPANY
☐ EMPLOYEE ☐ CONTRUCTOR		AM	PM			
☐ EMPLOYEE ☐ CONTRUCTOR		AM	PM			
☐ EMPLOYEE ☐ CONTRUCTOR		AM	PM			
☐ EMPLOYEE ☐ CONTRUCTOR		AM	PM			
☐ EMPLOYEE ☐ CONTRUCTOR		AM	PM			
☐ EMPLOYEE ☐ CONTRUCTOR		AM	PM			
☐ EMPLOYEE ☐ CONTRUCTOR		AM	PM			
☐ EMPLOYEE ☐ CONTRUCTOR		AM	PM			

EQUIPMENT ON SITE	NO. OF UNITE	WORKING YES / NO

HIRED EQUIPMENT	NO. OF UNITE	EQUIPMENT RENTED	FROM	RATE

NAME: _____   SIGNATURE: _____

MO TU WE TH FR SA SU
☐   ☐   ☐   ☐   ☐   ☐   ☐

DATE: ___/___/___

PROJECT:

FOREMAN:

WEATHER

F°_____ C°_____ _____AM _____PM

HOURS DUE TO BAD WEATHER	ISSUED AND DELAYS

NOTE: _____

COMPLETION DATE	DAYS AHEAD OF SCHEDULE	DAYS BEHIND SCHEDULE

SAFETY AND INCIDENTS

SAFETY ISSUES THAT NEED TO BE ADDRESSED	ACCIDENTS / INCIDENTS / STEPS NEEDED TO RESOLVE

SUMMARY OF THE WORK DONE TODAY

IMPORTANT NOTES

NAME	SIGNATURE

TODAY LABOR

INITIALS	TRADE	START	FINISH	PAID HOURS	OVERTIME	COMPANY
☐ EMPLOYEE ☐ CONTRUCTOR		AM	PM			
☐ EMPLOYEE ☐ CONTRUCTOR		AM	PM			
☐ EMPLOYEE ☐ CONTRUCTOR		AM	PM			
☐ EMPLOYEE ☐ CONTRUCTOR		AM	PM			
☐ EMPLOYEE ☐ CONTRUCTOR		AM	PM			
☐ EMPLOYEE ☐ CONTRUCTOR		AM	PM			
☐ EMPLOYEE ☐ CONTRUCTOR		AM	PM			
☐ EMPLOYEE ☐ CONTRUCTOR		AM	PM			

EQUIPMENT ON SITE	NO. OF UNITE	WORKING YES / NO

HIRED EQUIPMENT	NO. OF UNITE	EQUIPMENT RENTED	FROM	RATE

NAME: _____   SIGNATURE: _____

MO TU WE TH FR SA SU
☐   ☐   ☐   ☐   ☐   ☐   ☐

DATE: __ / __ / __

PROJECT:

FOREMAN:

WEATHER

F° ___ C° ___ ___ AM ___ PM

| HOURS DUE TO BAD WEATHER | ISSUED AND DELAYS |

NOTE: _____

COMPLETION DATE	DAYS AHEAD OF SCHEDULE	DAYS BEHIND SCHEDULE

SAFETY AND INCIDENTS

SAFETY ISSUES THAT NEED TO BE ADDRESSED	ACCIDENTS / INCIDENTS / STEPS NEEDED TO RESOLVE

SUMMARY OF THE WORK DONE TODAY

IMPORTANT NOTES

NAME	SIGNATURE

TODAY LABOR

INITIALS	TRADE	START	FINISH	PAID HOURS	OVERTIME	COMPANY
☐ EMPLOYEE ☐ CONTRUCTOR		AM	PM			
☐ EMPLOYEE ☐ CONTRUCTOR		AM	PM			
☐ EMPLOYEE ☐ CONTRUCTOR		AM	PM			
☐ EMPLOYEE ☐ CONTRUCTOR		AM	PM			
☐ EMPLOYEE ☐ CONTRUCTOR		AM	PM			
☐ EMPLOYEE ☐ CONTRUCTOR		AM	PM			
☐ EMPLOYEE ☐ CONTRUCTOR		AM	PM			
☐ EMPLOYEE ☐ CONTRUCTOR		AM	PM			

EQUIPMENT ON SITE	NO. OF UNITE	WORKING YES / NO

HIRED EQUIPMENT	NO. OF UNITE	EQUIPMENT RENTED	FROM	RATE

NAME: _____   SIGNATURE: _____

MO	TU	WE	TH	FR	SA	SU		DATE:	/	/
☐	☐	☐	☐	☐	☐	☐				

PROJECT: **FOREMAN:**

WEATHER

F°_____ C°_____ _____AM _____PM

HOURS DUE TO BAD WEATHER	ISSUED AND DELAYS

NOTE: _____

COMPLETION DATE	DAYS AHEAD OF SCHEDULE	DAYS BEHIND SCHEDULE

SAFETY AND INCIDENTS

SAFETY ISSUES THAT NEED TO BE ADDRESSED	ACCIDENTS / INCIDENTS / STEPS NEEDED TO RESOLVE

SUMMARY OF THE WORK DONE TODAY

IMPORTANT NOTES

NAME	SIGNATURE

TODAY LABOR

INITIALS	TRADE	START	FINISH	PAID HOURS	OVERTIME	COMPANY
☐ EMPLOYEE ☐ CONTRUCTOR		AM	PM			
☐ EMPLOYEE ☐ CONTRUCTOR		AM	PM			
☐ EMPLOYEE ☐ CONTRUCTOR		AM	PM			
☐ EMPLOYEE ☐ CONTRUCTOR		AM	PM			
☐ EMPLOYEE ☐ CONTRUCTOR		AM	PM			
☐ EMPLOYEE ☐ CONTRUCTOR		AM	PM			
☐ EMPLOYEE ☐ CONTRUCTOR		AM	PM			
☐ EMPLOYEE ☐ CONTRUCTOR		AM	PM			

EQUIPMENT ON SITE	NO. OF UNITE	WORKING YES / NO

HIRED EQUIPMENT	NO. OF UNITE	EQUIPMENT RENTED	FROM	RATE

NAME: _____   SIGNATURE: _____

MO TU WE TH FR SA SU

☐ ☐ ☐ ☐ ☐ ☐ ☐

DATE: ___/___/___

PROJECT:

FOREMAN:

WEATHER

F°___ C°___ ___ AM ___ PM

HOURS DUE TO BAD WEATHER

ISSUED AND DELAYS

NOTE: _____

COMPLETION DATE	DAYS AHEAD OF SCHEDULE	DAYS BEHIND SCHEDULE

SAFETY AND INCIDENTS

SAFETY ISSUES THAT NEED TO BE ADDRESSED	ACCIDENTS / INCIDENTS / STEPS NEEDED TO RESOLVE

SUMMARY OF THE WORK DONE TODAY

IMPORTANT NOTES

NAME	SIGNATURE

TODAY LABOR

INITIALS	TRADE	START	FINISH	PAID HOURS	OVERTIME	COMPANY
☐ EMPLOYEE ☐ CONTRUCTOR		AM	PM			
☐ EMPLOYEE ☐ CONTRUCTOR		AM	PM			
☐ EMPLOYEE ☐ CONTRUCTOR		AM	PM			
☐ EMPLOYEE ☐ CONTRUCTOR		AM	PM			
☐ EMPLOYEE ☐ CONTRUCTOR		AM	PM			
☐ EMPLOYEE ☐ CONTRUCTOR		AM	PM			
☐ EMPLOYEE ☐ CONTRUCTOR		AM	PM			
☐ EMPLOYEE ☐ CONTRUCTOR		AM	PM			

EQUIPMENT ON SITE	NO. OF UNITE	WORKING YES / NO

HIRED EQUIPMENT	NO. OF UNITE	EQUIPMENT RENTED	FROM	RATE

NAME: _____   SIGNATURE: _____

MO TU WE TH FR SA SU
☐   ☐   ☐   ☐   ☐   ☐   ☐

DATE: /   /

PROJECT:

FOREMAN:

WEATHER

F°_____ C°_____   _____AM _____PM

| HOURS DUE TO BAD WEATHER | ISSUED AND DELAYS |

NOTE: _____

COMPLETION DATE	DAYS AHEAD OF SCHEDULE	DAYS BEHIND SCHEDULE

SAFETY AND INCIDENTS

SAFETY ISSUES THAT NEED TO BE ADDRESSED	ACCIDENTS / INCIDENTS / STEPS NEEDED TO RESOLVE

SUMMARY OF THE WORK DONE TODAY

IMPORTANT NOTES

NAME	SIGNATURE

TODAY LABOR

INITIALS	TRADE	START	FINISH	PAID HOURS	OVERTIME	COMPANY
☐ EMPLOYEE ☐ CONTRUCTOR		AM	PM			
☐ EMPLOYEE ☐ CONTRUCTOR		AM	PM			
☐ EMPLOYEE ☐ CONTRUCTOR		AM	PM			
☐ EMPLOYEE ☐ CONTRUCTOR		AM	PM			
☐ EMPLOYEE ☐ CONTRUCTOR		AM	PM			
☐ EMPLOYEE ☐ CONTRUCTOR		AM	PM			
☐ EMPLOYEE ☐ CONTRUCTOR		AM	PM			
☐ EMPLOYEE ☐ CONTRUCTOR		AM	PM			

EQUIPMENT ON SITE	NO. OF UNITE	WORKING YES / NO

HIRED EQUIPMENT	NO. OF UNITE	EQUIPMENT RENTED	FROM	RATE

NAME: _____   SIGNATURE: _____

MO TU WE TH FR SA SU
☐   ☐   ☐   ☐   ☐   ☐   ☐

DATE: ___ / ___ / ___

PROJECT: _____

FOREMAN: _____

WEATHER

F° ___ C° ___ _____ AM _____ PM

HOURS DUE TO BAD WEATHER	ISSUED AND DELAYS

NOTE: _____

COMPLETION DATE	DAYS AHEAD OF SCHEDULE	DAYS BEHIND SCHEDULE

SAFETY AND INCIDENTS

SAFETY ISSUES THAT NEED TO BE ADDRESSED	ACCIDENTS / INCIDENTS / STEPS NEEDED TO RESOLVE

SUMMARY OF THE WORK DONE TODAY

IMPORTANT NOTES

NAME	SIGNATURE

TODAY LABOR

INITIALS	TRADE	START	FINISH	PAID HOURS	OVERTIME	COMPANY
☐ EMPLOYEE ☐ CONTRUCTOR		AM	PM			
☐ EMPLOYEE ☐ CONTRUCTOR		AM	PM			
☐ EMPLOYEE ☐ CONTRUCTOR		AM	PM			
☐ EMPLOYEE ☐ CONTRUCTOR		AM	PM			
☐ EMPLOYEE ☐ CONTRUCTOR		AM	PM			
☐ EMPLOYEE ☐ CONTRUCTOR		AM	PM			
☐ EMPLOYEE ☐ CONTRUCTOR		AM	PM			
☐ EMPLOYEE ☐ CONTRUCTOR		AM	PM			

EQUIPMENT ON SITE	NO. OF UNITE	WORKING YES / NO

HIRED EQUIPMENT	NO. OF UNITE	EQUIPMENT RENTED	FROM	RATE

NAME: _____   SIGNATURE: _____

MO   TU   WE   TH   FR   SA   SU   DATE: /   /
☐   ☐   ☐   ☐   ☐   ☐   ☐

PROJECT: FOREMAN:

WEATHER | HOURS DUE TO | ISSUED AND DELAYS
                                       | BAD WEATHER |
         F°___ C°___ ___AM ___PM | |

NOTE: _____

COMPLETION DATE	DAYS AHEAD OF SCHEDULE	DAYS BEHIND SCHEDULE

SAFETY AND INCIDENTS

SAFETY ISSUES THAT NEED TO BE ADDRESSED	ACCIDENTS / INCIDENTS / STEPS NEEDED TO RESOLVE

SUMMARY OF THE WORK DONE TODAY

IMPORTANT NOTES

NAME	SIGNATURE

TODAY LABOR

INITIALS	TRADE	START	FINISH	PAID HOURS	OVERTIME	COMPANY
☐ EMPLOYEE ☐ CONTRUCTOR		AM	PM			
☐ EMPLOYEE ☐ CONTRUCTOR		AM	PM			
☐ EMPLOYEE ☐ CONTRUCTOR		AM	PM			
☐ EMPLOYEE ☐ CONTRUCTOR		AM	PM			
☐ EMPLOYEE ☐ CONTRUCTOR		AM	PM			
☐ EMPLOYEE ☐ CONTRUCTOR		AM	PM			
☐ EMPLOYEE ☐ CONTRUCTOR		AM	PM			
☐ EMPLOYEE ☐ CONTRUCTOR		AM	PM			

EQUIPMENT ON SITE	NO. OF UNITE	WORKING YES / NO

HIRED EQUIPMENT	NO. OF UNITE	EQUIPMENT RENTED	FROM	RATE

NAME: _____   SIGNATURE: _____

MO TU WE TH FR SA SU
☐   ☐   ☐   ☐   ☐   ☐   ☐

DATE: ___ / ___ / ___

PROJECT: _____

FOREMAN: _____

WEATHER ☁ ⛅ ☁ 🌨 ☀ 🌧 ⛈

F°___ C°___ ___AM ___PM

HOURS DUE TO BAD WEATHER	ISSUED AND DELAYS

NOTE: _____

COMPLETION DATE	DAYS AHEAD OF SCHEDULE	DAYS BEHIND SCHEDULE

SAFETY AND INCIDENTS

SAFETY ISSUES THAT NEED TO BE ADDRESSED	ACCIDENTS / INCIDENTS / STEPS NEEDED TO RESOLVE

SUMMARY OF THE WORK DONE TODAY

IMPORTANT NOTES

NAME	SIGNATURE

TODAY LABOR

INITIALS	TRADE	START	FINISH	PAID HOURS	OVERTIME	COMPANY
☐ EMPLOYEE ☐ CONTRUCTOR		AM	PM			
☐ EMPLOYEE ☐ CONTRUCTOR		AM	PM			
☐ EMPLOYEE ☐ CONTRUCTOR		AM	PM			
☐ EMPLOYEE ☐ CONTRUCTOR		AM	PM			
☐ EMPLOYEE ☐ CONTRUCTOR		AM	PM			
☐ EMPLOYEE ☐ CONTRUCTOR		AM	PM			
☐ EMPLOYEE ☐ CONTRUCTOR		AM	PM			
☐ EMPLOYEE ☐ CONTRUCTOR		AM	PM			

EQUIPMENT ON SITE	NO. OF UNITE	WORKING YES / NO

HIRED EQUIPMENT	NO. OF UNITE	EQUIPMENT RENTED	FROM	RATE

NAME: _____   SIGNATURE: _____

MO TU WE TH FR SA SU
☐   ☐   ☐   ☐   ☐   ☐   ☐

DATE: ___ / ___ / ___

PROJECT: _____

FOREMAN: _____

WEATHER

F° ___ C° ___   ___ AM ___ PM

HOURS DUE TO BAD WEATHER	ISSUED AND DELAYS

NOTE: _____

COMPLETION DATE	DAYS AHEAD OF SCHEDULE	DAYS BEHIND SCHEDULE

SAFETY AND INCIDENTS

SAFETY ISSUES THAT NEED TO BE ADDRESSED	ACCIDENTS / INCIDENTS / STEPS NEEDED TO RESOLVE

SUMMARY OF THE WORK DONE TODAY

IMPORTANT NOTES

NAME	SIGNATURE

TODAY LABOR

INITIALS	TRADE	START	FINISH	PAID HOURS	OVERTIME	COMPANY
☐ EMPLOYEE ☐ CONTRUCTOR		AM	PM			
☐ EMPLOYEE ☐ CONTRUCTOR		AM	PM			
☐ EMPLOYEE ☐ CONTRUCTOR		AM	PM			
☐ EMPLOYEE ☐ CONTRUCTOR		AM	PM			
☐ EMPLOYEE ☐ CONTRUCTOR		AM	PM			
☐ EMPLOYEE ☐ CONTRUCTOR		AM	PM			
☐ EMPLOYEE ☐ CONTRUCTOR		AM	PM			
☐ EMPLOYEE ☐ CONTRUCTOR		AM	PM			

EQUIPMENT ON SITE	NO. OF UNITE	WORKING YES / NO

HIRED EQUIPMENT	NO. OF UNITE	EQUIPMENT RENTED	FROM	RATE

NAME: _____ SIGNATURE: _____

MO TU WE TH FR SA SU

□   □   □   □   □   □   □   DATE: __/__/__

PROJECT: _____   FOREMAN: _____

WEATHER   HOURS DUE TO   ISSUED AND DELAYS
                                       BAD WEATHER
        F°___ C°___ ___ AM ___ PM

NOTE: _____

COMPLETION DATE	DAYS AHEAD OF SCHEDULE	DAYS BEHIND SCHEDULE

SAFETY AND INCIDENTS

SAFETY ISSUES THAT NEED TO BE ADDRESSED	ACCIDENTS / INCIDENTS / STEPS NEEDED TO RESOLVE
_____	_____
_____	_____
_____	_____
_____	_____

SUMMARY OF THE WORK DONE TODAY

_____
_____
_____
_____
_____

IMPORTANT NOTES

_____
_____
_____
_____
_____
_____

NAME	SIGNATURE

TODAY LABOR

INITIALS	TRADE	START	FINISH	PAID HOURS	OVERTIME	COMPANY
☐ EMPLOYEE ☐ CONTRUCTOR		AM	PM			
☐ EMPLOYEE ☐ CONTRUCTOR		AM	PM			
☐ EMPLOYEE ☐ CONTRUCTOR		AM	PM			
☐ EMPLOYEE ☐ CONTRUCTOR		AM	PM			
☐ EMPLOYEE ☐ CONTRUCTOR		AM	PM			
☐ EMPLOYEE ☐ CONTRUCTOR		AM	PM			
☐ EMPLOYEE ☐ CONTRUCTOR		AM	PM			
☐ EMPLOYEE ☐ CONTRUCTOR		AM	PM			

EQUIPMENT ON SITE	NO. OF UNITE	WORKING YES / NO

HIRED EQUIPMENT	NO. OF UNITE	EQUIPMENT RENTED	FROM	RATE

NAME: _____   SIGNATURE: _____

MO TU WE TH FR SA SU
☐   ☐   ☐   ☐   ☐   ☐   ☐

DATE: ___/___/___

PROJECT:

FOREMAN:

WEATHER

F°_____ C°_____ _____AM _____PM

HOURS DUE TO BAD WEATHER	ISSUED AND DELAYS

NOTE: _____

COMPLETION DATE	DAYS AHEAD OF SCHEDULE	DAYS BEHIND SCHEDULE

SAFETY AND INCIDENTS

SAFETY ISSUES THAT NEED TO BE ADDRESSED	ACCIDENTS / INCIDENTS / STEPS NEEDED TO RESOLVE

SUMMARY OF THE WORK DONE TODAY

IMPORTANT NOTES

NAME	SIGNATURE

TODAY LABOR

INITIALS	TRADE	START	FINISH	PAID HOURS	OVERTIME	COMPANY
☐ EMPLOYEE ☐ CONTRUCTOR		AM	PM			
☐ EMPLOYEE ☐ CONTRUCTOR		AM	PM			
☐ EMPLOYEE ☐ CONTRUCTOR		AM	PM			
☐ EMPLOYEE ☐ CONTRUCTOR		AM	PM			
☐ EMPLOYEE ☐ CONTRUCTOR		AM	PM			
☐ EMPLOYEE ☐ CONTRUCTOR		AM	PM			
☐ EMPLOYEE ☐ CONTRUCTOR		AM	PM			
☐ EMPLOYEE ☐ CONTRUCTOR		AM	PM			

EQUIPMENT ON SITE	NO. OF UNITE	WORKING YES / NO

HIRED EQUIPMENT	NO. OF UNITE	EQUIPMENT RENTED	FROM	RATE

NAME: _____   SIGNATURE: _____

MO TU WE TH FR SA SU
☐   ☐   ☐   ☐   ☐   ☐   ☐

DATE: __ / __ / __

PROJECT: _____

FOREMAN: _____

WEATHER ☁☂ ⛅ ☁ 🌦 ☀ 🌧 ⛈

F° ____ C° ____ ____ AM ____ PM

HOURS DUE TO BAD WEATHER	ISSUED AND DELAYS

NOTE: _____

COMPLETION DATE	DAYS AHEAD OF SCHEDULE	DAYS BEHIND SCHEDULE

SAFETY AND INCIDENTS

SAFETY ISSUES THAT NEED TO BE ADDRESSED	ACCIDENTS / INCIDENTS / STEPS NEEDED TO RESOLVE

SUMMARY OF THE WORK DONE TODAY

_____
_____
_____
_____
_____

IMPORTANT NOTES

_____
_____
_____
_____
_____

NAME	SIGNATURE

TODAY LABOR

INITIALS	TRADE	START	FINISH	PAID HOURS	OVERTIME	COMPANY
☐ EMPLOYEE ☐ CONTRUCTOR		AM	PM			
☐ EMPLOYEE ☐ CONTRUCTOR		AM	PM			
☐ EMPLOYEE ☐ CONTRUCTOR		AM	PM			
☐ EMPLOYEE ☐ CONTRUCTOR		AM	PM			
☐ EMPLOYEE ☐ CONTRUCTOR		AM	PM			
☐ EMPLOYEE ☐ CONTRUCTOR		AM	PM			
☐ EMPLOYEE ☐ CONTRUCTOR		AM	PM			
☐ EMPLOYEE ☐ CONTRUCTOR		AM	PM			

EQUIPMENT ON SITE	NO. OF UNITE	WORKING YES / NO

HIRED EQUIPMENT	NO. OF UNITE	EQUIPMENT RENTED	FROM	RATE

NAME: _____   SIGNATURE: _____

MO	TU	WE	TH	FR	SA	SU
☐	☐	☐	☐	☐	☐	☐

DATE: ___ / ___ / ___

PROJECT: _____

FOREMAN: _____

WEATHER

F° ____ C° ____ ____ AM ____ PM

HOURS DUE TO BAD WEATHER	ISSUED AND DELAYS

NOTE: _____

COMPLETION DATE	DAYS AHEAD OF SCHEDULE	DAYS BEHIND SCHEDULE

SAFETY AND INCIDENTS

SAFETY ISSUES THAT NEED TO BE ADDRESSED	ACCIDENTS / INCIDENTS / STEPS NEEDED TO RESOLVE

SUMMARY OF THE WORK DONE TODAY

IMPORTANT NOTES

NAME	SIGNATURE

TODAY LABOR

INITIALS	TRADE	START	FINISH	PAID HOURS	OVERTIME	COMPANY
☐ EMPLOYEE ☐ CONTRUCTOR		AM	PM			
☐ EMPLOYEE ☐ CONTRUCTOR		AM	PM			
☐ EMPLOYEE ☐ CONTRUCTOR		AM	PM			
☐ EMPLOYEE ☐ CONTRUCTOR		AM	PM			
☐ EMPLOYEE ☐ CONTRUCTOR		AM	PM			
☐ EMPLOYEE ☐ CONTRUCTOR		AM	PM			
☐ EMPLOYEE ☐ CONTRUCTOR		AM	PM			
☐ EMPLOYEE ☐ CONTRUCTOR		AM	PM			

EQUIPMENT ON SITE	NO. OF UNITE	WORKING YES / NO

HIRED EQUIPMENT	NO. OF UNITE	EQUIPMENT RENTED	FROM	RATE

NAME: _____   SIGNATURE: _____

MO TU WE TH FR SA SU
☐   ☐   ☐   ☐   ☐   ☐   ☐

DATE: ___/___/___

PROJECT: _____

FOREMAN: _____

WEATHER ☁️ 🌤️ ☁️ 🌨️ ☀️ 🌧️ ⛈️

F°____ C°____ ____AM ____PM

HOURS DUE TO BAD WEATHER	ISSUED AND DELAYS

NOTE: _____

COMPLETION DATE	DAYS AHEAD OF SCHEDULE	DAYS BEHIND SCHEDULE

SAFETY AND INCIDENTS

SAFETY ISSUES THAT NEED TO BE ADDRESSED	ACCIDENTS / INCIDENTS / STEPS NEEDED TO RESOLVE

SUMMARY OF THE WORK DONE TODAY

IMPORTANT NOTES

NAME	SIGNATURE

TODAY LABOR

INITIALS	TRADE	START	FINISH	PAID HOURS	OVERTIME	COMPANY
☐ EMPLOYEE ☐ CONTRUCTOR		AM	PM			
☐ EMPLOYEE ☐ CONTRUCTOR		AM	PM			
☐ EMPLOYEE ☐ CONTRUCTOR		AM	PM			
☐ EMPLOYEE ☐ CONTRUCTOR		AM	PM			
☐ EMPLOYEE ☐ CONTRUCTOR		AM	PM			
☐ EMPLOYEE ☐ CONTRUCTOR		AM	PM			
☐ EMPLOYEE ☐ CONTRUCTOR		AM	PM			
☐ EMPLOYEE ☐ CONTRUCTOR		AM	PM			

EQUIPMENT ON SITE	NO. OF UNITE	WORKING YES / NO

HIRED EQUIPMENT	NO. OF UNITE	EQUIPMENT RENTED	FROM	RATE

NAME: _____   SIGNATURE: _____

MO TU WE TH FR SA SU
☐   ☐   ☐   ☐   ☐   ☐   ☐

DATE: / /

PROJECT:

FOREMAN:

WEATHER

F°____ C°____ ____AM ____PM

HOURS DUE TO BAD WEATHER	ISSUED AND DELAYS

NOTE: _____

COMPLETION DATE	DAYS AHEAD OF SCHEDULE	DAYS BEHIND SCHEDULE

SAFETY AND INCIDENTS

SAFETY ISSUES THAT NEED TO BE ADDRESSED	ACCIDENTS / INCIDENTS / STEPS NEEDED TO RESOLVE

SUMMARY OF THE WORK DONE TODAY

IMPORTANT NOTES

NAME	SIGNATURE

TODAY LABOR

INITIALS	TRADE	START	FINISH	PAID HOURS	OVERTIME	COMPANY
☐ EMPLOYEE ☐ CONTRUCTOR		AM	PM			
☐ EMPLOYEE ☐ CONTRUCTOR		AM	PM			
☐ EMPLOYEE ☐ CONTRUCTOR		AM	PM			
☐ EMPLOYEE ☐ CONTRUCTOR		AM	PM			
☐ EMPLOYEE ☐ CONTRUCTOR		AM	PM			
☐ EMPLOYEE ☐ CONTRUCTOR		AM	PM			
☐ EMPLOYEE ☐ CONTRUCTOR		AM	PM			
☐ EMPLOYEE ☐ CONTRUCTOR		AM	PM			

EQUIPMENT ON SITE	NO. OF UNITE	WORKING YES / NO

HIRED EQUIPMENT	NO. OF UNITE	EQUIPMENT RENTED	FROM	RATE

NAME: _____   SIGNATURE: _____

MO TU WE TH FR SA SU
☐   ☐   ☐   ☐   ☐   ☐   ☐

DATE: ____ / ____ / ____

PROJECT: _____

FOREMAN: _____

WEATHER   F°____ C°____   ____AM   ____PM

HOURS DUE TO BAD WEATHER	ISSUED AND DELAYS

NOTE: _____

COMPLETION DATE	DAYS AHEAD OF SCHEDULE	DAYS BEHIND SCHEDULE

SAFETY AND INCIDENTS

SAFETY ISSUES THAT NEED TO BE ADDRESSED	ACCIDENTS / INCIDENTS / STEPS NEEDED TO RESOLVE

SUMMARY OF THE WORK DONE TODAY

IMPORTANT NOTES

NAME	SIGNATURE

TODAY LABOR

INITIALS	TRADE	START	FINISH	PAID HOURS	OVERTIME	COMPANY
☐ EMPLOYEE ☐ CONTRUCTOR		AM	PM			
☐ EMPLOYEE ☐ CONTRUCTOR		AM	PM			
☐ EMPLOYEE ☐ CONTRUCTOR		AM	PM			
☐ EMPLOYEE ☐ CONTRUCTOR		AM	PM			
☐ EMPLOYEE ☐ CONTRUCTOR		AM	PM			
☐ EMPLOYEE ☐ CONTRUCTOR		AM	PM			
☐ EMPLOYEE ☐ CONTRUCTOR		AM	PM			
☐ EMPLOYEE ☐ CONTRUCTOR		AM	PM			

EQUIPMENT ON SITE	NO. OF UNITE	WORKING YES / NO

HIRED EQUIPMENT	NO. OF UNITE	EQUIPMENT RENTED	FROM	RATE

NAME: _____   SIGNATURE: _____

MO TU WE TH FR SA SU
☐   ☐   ☐   ☐   ☐   ☐   ☐

DATE: ___ / ___ / ___

PROJECT:

FOREMAN:

WEATHER

F° ___ C° ___ ___ AM ___ PM

| HOURS DUE TO BAD WEATHER | ISSUED AND DELAYS |

NOTE: _____

COMPLETION DATE	DAYS AHEAD OF SCHEDULE	DAYS BEHIND SCHEDULE

SAFETY AND INCIDENTS

SAFETY ISSUES THAT NEED TO BE ADDRESSED	ACCIDENTS / INCIDENTS / STEPS NEEDED TO RESOLVE

SUMMARY OF THE WORK DONE TODAY

IMPORTANT NOTES

NAME	SIGNATURE

TODAY LABOR

INITIALS	TRADE	START	FINISH	PAID HOURS	OVERTIME	COMPANY
☐ EMPLOYEE ☐ CONTRUCTOR		AM	PM			
☐ EMPLOYEE ☐ CONTRUCTOR		AM	PM			
☐ EMPLOYEE ☐ CONTRUCTOR		AM	PM			
☐ EMPLOYEE ☐ CONTRUCTOR		AM	PM			
☐ EMPLOYEE ☐ CONTRUCTOR		AM	PM			
☐ EMPLOYEE ☐ CONTRUCTOR		AM	PM			
☐ EMPLOYEE ☐ CONTRUCTOR		AM	PM			
☐ EMPLOYEE ☐ CONTRUCTOR		AM	PM			

EQUIPMENT ON SITE	NO. OF UNITE	WORKING YES / NO

HIRED EQUIPMENT	NO. OF UNITE	EQUIPMENT RENTED	FROM	RATE

NAME: _____   SIGNATURE: _____

IMPORTANT TELEPHONE NUMBER

▶ NAME _____ PHONE _____

EMAIL _____

▶ NAME _____ PHONE _____

EMAIL _____

▶ NAME _____ PHONE _____

EMAIL _____

▶ NAME _____ PHONE _____

EMAIL _____

▶ NAME _____ PHONE _____

EMAIL _____

▶ NAME _____ PHONE _____

EMAIL _____

▶ NAME _____ PHONE _____

EMAIL _____

▶ NAME _____ PHONE _____

EMAIL _____

▶ NAME _____ PHONE _____

EMAIL _____

▶ NAME _____ PHONE _____

EMAIL _____

▶ NAME _____ PHONE _____

EMAIL _____

▶ NAME _____ PHONE _____

EMAIL _____

▶ NAME _____ PHONE _____

EMAIL _____